I0067021

Ernst McConkey

Cyclopedia of medicine

The household friend

Ernst McConkey

Cyclopedia of medicine
The household friend

ISBN/EAN: 9783337223724

Printed in Europe, USA, Canada, Australia, Japan

Cover: Foto ©berggeist007 / pixelio.de

More available books at **www.hansebooks.com**

CYCLOPEDIA OF MEDICINE

'R THE

Household Friend;

CONTAINING

VALUABLE PRESCRIPTIONS AND FORMULÆ

THAT HAVE BEEN SUCCESSFULLY USED BY THE
LEADING PHYSICIANS AND SPECIALISTS
OF EUROPE AND AMERICA.

PRESCRIPTIONS AND DIRECTIONS IN PLAIN ENGLISH SO THAT
ANYONE CAN UNDERSTAND THEM.

PRICE, $2.00.

COMPILED AND PUBLISHED BY

PROFESSOR E. McCONKEY.

CHICAGO, ILL.
1892.

THESE Prescriptions have been secured at great expense from noted physicians and specialists, who are using them successfully in their practice, and have all been thoroughly tested.

Having traveled throughout the United States for twenty-five years in the medicine business, and being a close observer, I have found that thousands die every year for the want of treatment at the proper time.

Recipes Nos. 1, 2, and 3 should always be kept in the house, to be used in cases of emergency.

Knowing that this book will be found a friend in time of need, it is offered to the public.

Prof. E. McConkey.

IN buying this book, I promise on my sacred word of honor not to allow anyone outside of my own family to read its contents or to copy therefrom.

CONTENTS.

No. 1

General Family Remedy.

The best medicine known today.

℞ Alcohol, - - - - 8 ounces.
Oil Origanum, - - - 1 dram.
Oil Sassafras, - - 1 dram.
Oil Hemlock, - - - 1 dram.
Oil Cloves, - - - 1 dram.
Tinct. Opium, - - - 1 dram.
Tinct. Capsicum, - - 2 drams.
Sulph. Ether, - - - 6 drams.
Aqua Ammonia, - - 2 drams.
Gum Camphor, - - - 1 dram.

DOSE:—For an adult, one teaspoonful in a wine-glass of water. When given to children, reduce the dose according to the age of the child.

For Rheumatism, Neuralgia, Lame Back, Inflammation of the Kidneys, take internally three times a day and bathe externally with the clear medicine.

For Toothache, bathe each side of the gums; Headache, bathe freely; Cramp Colic, take a dose in hot water every half-hour.

For Bites or Stings of Insects, Reptiles, or Dogs, apply the clear medicine.

For Sore Throat, mix a teaspoonful in a wine-glass of water, and gargle the throat and bathe on the outside several times a day.

No. 2
Cholera Mixture.
FOR CHOLERA MORBUS OR SUMMER COMPLAINT.

Is a positive cure and may truthfully be said that in twenty years it has never been known to fail in a single case.

R Spts. Lavender, Comp. - 2 ounces.
Tinct. Rhubarb, - - - 2 ounces.
Tinct. Opium, - - 2 ounces.
Tinct. Camphor, - - 2 ounces.
Tinct. Catechu, - - 1 ounce.
Syrup Simple, - - - 1 ounce.

DOSE:—For an adult, one teaspoonful; for a child under 12 years, one-half teaspoonful; for a child under 2 years, ten drops. To be given in a little sweetened water.

No. 3
Scrofula.

This remedy has been used with perfect success after several physicians had failed.

R Syrup of Stillingia, - - 2 ounces.
Syrup of Sarsaparilla, - 6 ounces.
Iodide Potassium, - - 3 drams.

DIRECTIONS:—Take one teaspoonful before meals.

No. 4

Alcoholism.

SURE CURE FOR DEPSOMANIA OR DRUNKENNESS.

This is a positive cure, if properly put up, as thousands can testify today.

℞ Take one pound of Red Cinchona Bark; use the quill; crush it in a mortar; it does not want to be pulverized fine, simply crushed into small pieces. Put it into a percolator and pour about a half-pint of diluted Alcohol onto it, and let it stand two or three days to become thoroughly soaked. Then gradually pour on more diluted Alcohol until you have one pint of the tincture run through the percolator. Then take this one pint of tincture and evaporate over a slow fire until reduced to one half-pint.

DOSE:—One teaspoonful three or four times a day. If at any time it should produce a ringing sound in the head, reduce the size of the dose.

No. 5

Constipation.

LAXATIVE AND SURE CURE FOR CONSTIPATION.

℞ Fl. Ext. Belladonna, - ½ dram.
Fl. Ext. Cascara Sagrada, - 1 ounce.
Syrup Rhubarb, - - 3 ounces.

DOSE:—Teaspoonful every three hours until action, then take as needed.

No. 6

Fever-and-Ague Cure.

FOR AGUE, CHILLS, AND FEVER.

℞ Salicylate Soda,   -   -   1½ drams.
    Sulphate Quinine,   -   -   1½ drams.
    Syrup of Rhubarb,   -   6 ounces.

DOSE:—Teaspoonful three times a day. Keep the patient in bed the days they expect a chill and give double doses every two hours until you give three doses, commencing five hours before time for the chill.

No. 7

Kidneys.

For the Kidneys when there is any deposit or sediment in the urine.

℞ Nitrate of Potassium,   -   3 drams.
    Pulv. Cubebs,   -   -   2 drams.
    Rectified Spirits,   -   -   1 dram.
    Water sufficient to make -   3 ounces.

DOSE:—Teaspoonful three times a day, in water.

No. 8

Chronic Dysentery Pills.

℞ Peptonic Pills,   -   -   -   No. xxx

DOSE:—One after each meal, as needed.

No. 9
Blood Purifier.

No better Blood Purifier known. Recommended by physicians, and superior to any medicine advertised.

R Fl. Ext. Cascara Sagrada, 4 drams.
 Fl. Ext. Sarsaparilla, - - 4 drams.
 Fl. Ext. Stillingia, - - 4 drams.
 Iodide Potash, - - - 4 drams.
 Simple Syrup, to make 8 ounces.

DOSE:—Teaspoonful three times a day.

No. 10
Cramp-Colic.

COLIC, CRAMPS, AND GRIPING PAINS IN THE BOWELS.

R Tinct. Capsicum, - - 4 drams.
 Tinct. Camphor, - - - 4 drams.
 Sulphuric Ether, - - 4 drams.
 Spirits Peppermint, - - 4 drams.
 Glycerine, - - - 1 ounce.

DOSE:—One-half to one teaspoonful in a little water every hour till relieved. Relief will follow immediately by using this remedy.

No. 11
Female Tonic.

℞ Elixir Beef, Wine and Iron,   6 ounces.
Carbonate of Iron,   -   - 4 drams.
Elixir Calisya Bark,   -   4 ounces.
Simple Syrup, -   -   - 2 ounces.

DOSE:—Teaspoonful to tablespoonful three times a day.

No. 12
Catarrh Salve.

℞ Bi-carb Soda, -   -   -   1 dram.
Cosmoline, -   -   -   - 1 ounce.
Mix thoroughly.

DIRECTIONS:—Place a little of the salve on the end of the little finger and grease the inner nostrils, and rub on the bridge of the nose and between the eyes on retiring for the night.

No. 13
Coughs.

℞ Ammonia Chloride,   -   4 drams.
Prussic Acid, diluted,   16 drops.
Tinct. Henbane,   -   -   2 drams.
Hive-Syrup Comp., -   - 6 drams.
Syrup Licorice,   -   -   5 ounces.

DOSE:—Tablespoonful every three hours for adults.

Cough Mixture.

FOR COUGHS, COLDS, AND PULMONARY COMPLAINTS.

This is the most wonderful cough remedy known, and has cured several cases where they were supposed to have consumption.

℞ Hops, - - - - 1 ounce.
Wild-Cherry Bark, - - 1 ounce.
Spiknard, - - - 1 ounce.
Elecampane, - - - 4 drams.

Put the herbs into three pints of water and simmer down one-third; do not boil. Strain through a cloth; add one pound of loaf sugar, then take quarter of a teaspoonful of tar on a stick and stir mixture. To preserve, add quarter of a pint of alcohol.

DOSE:—Tablespoonful three times a day.

To Break up a Cold.

℞ Dovers Powder, - - 20 grains.
Nitrate Potash, - - 20 grains.
Camphor, - - - 10 grains.
Pulv. Cubebs, - - - 20 grains.

DIRECTIONS:—Mix and divide into ten capsules or powders, and take one every six hours.

No. 16

Whooping-Cough.

This is superior to all remedies in the market.

℞ Onions and Garlic, bruised, each 4 ounces.
Sweet Oil, - - - 4 ounces.
Stew and strain.
Then add Honey, - - - 4 ounces.
Paregoric, - - · - - 4 drams.
Spirits Camphor, - - - 4 drams.

DOSE:—Teaspoonful three or four times a day.

No. 17

Piles.

SURE CURE FOR PILES.

Have known this to cure hundreds, and never knew it to fail.

℞ Acetate of Lead, - - 1 dram.
Ext. Belladonna, - - 1 dram.
Tanic Acid, - - - 1 dram.
Opium, Pulv. - - - ½ dram.
Vaseline, - - - 1 ounce.

DIRECTIONS:—Apply night and morning.

No. 18

Rheumatism.

SURE CURE FOR RHEUMATISM.

℞ Tr. Colchicum Seed,   -   1 ounce.
Soda Salicylate,   -   -   1 ounce.
Fl. Ext. Black Cohosh,   -   1 ounce.
Fl. Ext. Poke Root,   -   -   1 ounce.
Syrup Rhubarb,   -   -   6 ounces.

DIRECTIONS:—Teaspoonful in water every 4 hours.
Apply No. 1 externally three or four times a day.

No. 19

Chronic Rheumatism.

℞ Bi-carb. Potassa,   -   -   1 ounce.
Salicylate Soda,   -   -   1 ounce.
Potassi Iodidi,   -   -   2 drams.
Water sufficient to make   6 ounces.

Filter solution.

DIRECTIONS:—Teaspoonful in a little water three
times daily before meals. On going to bed, take a
quarter-grain gelatine-coated Podophyllin Pill.

No. 20
Headache Cure.

FOR HEADACHE, PAIN IN THE BONES, BREAKBONE
FEVER, OR DENGUE FEVER.

℞ Phenacetine. Have powders made containing seven grains each. Take a powder every five or six hours until relieved; should be taken in a little wine, but can be taken dry and washed down with water.

No. 21
La Grippe.

Recommended by the leading physicians of the South.

℞ Phenacetine and Quinine.

DIRECTIONS:—Have capsules made containing $2\frac{1}{2}$ grains of Quinine and $3\frac{1}{2}$ grains of Phenacetine. Take two capsules every five hours until relieved.

No. 22
Sore Eyes.

℞				
Sulphate Zinc,	-	-	2 grains.
Acid Boracic,	-	-	-	5 grains.
Morphine,	-	-	-	$\frac{1}{4}$ grain.
Rose-Water,		-	-	1 ounce.

DIRECTIONS:—Drop a few drops in the eye, and bathe freely once a day.

No. 23

Mouth Wash.

For Ulcerated and Inflamed Gums, Sore Mouth, Offensive Breath, and Preservation of Teeth.

℞ Acid Carb. Scherings,	-	2 drams.
Comp. Spirits Lavender,	-	1½ ounces.
Tinct. Myrrh, -	- -	1½ ounces.
Oil Wintergreen,	- -	45 drops.
Oil Peppermint,	- -	30 drops.
Oil Cloves,	- - -	20 drops.
Pulv. Orris Florentine,	-	2 drams.
Soda Bi-carbonate,	- -	2 drams.
White Sugar, -	- -	1 ounce.
Alcohol,	- - - -	4 ounces.
Aqua,	- - - -	11 ounces.

Macerate 24 hours and filter.

DIRECTIONS:—For Toilet Uses, put a few drops on brush night and morning, thereby imparting a delightful and refreshing taste to the mouth.

For Sore Gums, etc., put half a teaspoonful in two tablespoonfuls of water, and rinse the mouth every two or three hours.

No. 24
Worms.
Vermifuge for Worms in Children.

R   Pulverized Santonin,   -   1 dram.
    Fl. Ext. Cascara Sagrada, - 1 ounce.
    Syrup Rhubarb,   -   -   2 ounces.
    Essence Peppermint, -   -   1 dram.

DIRECTIONS:—Teaspoonful night and morning till worms are expelled.

No. 25
Great German Salve.

Has no equal on earth; acknowledged to be the best salve in use today by thousands who have used it.

R   White Wax,   -   -   -   1 ounce.
    Gum Camphor,   -   -   -   1 ounce.
    Best Sweet Oil,   -   -   1 ounce.
    Precipitate Powder,   -   -   1 ounce.

Add a piece of fresh Butter the size of an egg; melt all together except the powder. When all is dissolved, put in the powder, and keep stirring until cool enough to prevent the powder from settling. Keep in air-tight box.

DIRECTIONS:—First clean the sore with Castile Soap; if there is any proud flesh, sprinkled on pulverized burnt alum. Apply on soft cloth twice a day.

No. 26

Dyspepsia.

R Carbonate Magnesia,   -   4 drams.
   Bi-carbonate Soda,   -   - 1 ounce.
   Subnitrate Bismuth,   -   4 drams.
   Sugar of Peppermint,   -   1 ounce.

Dose:—Teaspoonful in water three times a day, before or after eating.

No. 27

Nervousness.

R Fluid Extract Celery Seed,   1 ounce.
   Fluid Extract Chamomile, -   1 ounce.
   Fluid Extract Hyoscyamus,   4 drams.
   Fluid Extract Licorice,   -   4 drams.
   Simple Elixir,   -   -   13 ounces.

Filter sig.

Dose:—Take a teaspoonful three times a day: at 9 o'clock in the morning, 3 o'clock in the afternoon, and 9 o'clock at night.

No. 28

Summer Complaint.

FOR SUMMER COMPLAINT AND ALL BOWEL TROUBLE
WITH CHILDREN.

℞ Rhubarb Root,	-	-	4 drams.
Anise Seed,	-	-	4 drams.
Licorice Root,	-	-	4 drams.
Manna,	-	-	4 drams.
Powdered Sugar,	-	-	1 pound.
Paregoric,	-	-	1 ounce.
Carbonate of Potash,	-		80 grains.

DIRECTIONS:—Simmer the Rhubarb, Anise, and
Licorice with 16 ounces of water till reduced to 12
ounces; then add Manna and strain. Make a syrup
and add Carbonate of Potash and 3 ounces of French
Brandy.

DOSE:—Teaspoonful three or four times a day for
an infant; for children from four to ten years old,
two teaspoonfuls; for adults, from one to two tablespoonfuls.

No. 29

Stomach, Liver, and Kidneys.

℞ Fluid Extract Golden-seal,			3 drams.
Tincture Cubebs,	-	-	1 ounce.
Nitrate of Potash,	-	-	4 drams.
Water to make	-	-	6 ounces.

DOSE:—A teaspoonful in a tablespoonful of water
three times a day: 9 a.m., 3 p.m. and 9 p.m.

No. 30

Liver, Kidneys, and Blood.

A great system renovator. Regulates the Liver and Kidneys, purifies the Blood, and is a splendid tonic.

R Carbonate of Iron, - - 4 drams.
Pulverized Cubebs, - 2 drams.
Tincture Aloes, - - - 2 ounces.
Alcohol, - - - 6 drams.
Soft water to make - - 6 ounces.

DOSE:—A teaspoonful two or three times a day.

No. 31

For the Liver.

To be used in all cases instead of Calomel.

R Irisin, - - - - 24 grains.
Divide into three powders.

DIRECTIONS:—Take one powder at night on going to bed. In the morning take Senna Tea or Seidlitz Powder.

Try it and you will use no more Calomel.

No. 32

Tape-Worm Remedy.

WILL NEVER FAIL IN EXPELLING TAPE-WORM.

℞ Take one pint of Pumpkin Seed, soak them in water twenty minutes, then pour off the water. Then add one quart of water to the seed and simmer very slowly down to one pint; put all into a cloth and wring dry; set the tea one side and hull some of the seed you have been using; keep the hulled seed and tea until morning.

DIRECTIONS:—The day the tea is made eat a very light breakfast, and do not eat any more during the day. As soon as you get up in the morning take a good drink of the tea and eat as much seed as you can, thus satisfying your hunger for breakfast. Drink all the tea within an hour after getting up; two hours after taking the last of the tea, take a large dose of Castor Oil, to move the bowels.

No. 33

Lucorrhea.

Injection for Lucorrhea or Falling of the Womb.

℞ Golden-Seal Powder, - 1 ounce.
Alum Powder, - - - 4 ounces.
Borax Powder, - - 2 ounces.

DIRECTIONS:—Take one teaspoonful and dissolve in a pint of hot water, and inject night and morning.

No. 34

Gonorrhœa.

Internal Remedy.

℞ Tincture of Cubebs,   -   1 ounce.
    Balsam Copaiba,   -   -   1 ounce.
    Spirits of Nitre,   -   -   1 ounce.
    Syrup Tolu,   -   -   -   3 ounces.

DOSE:—Teaspoonful three times a day: at 9 a.m., 3 p.m., and 9 p.m.

No. 35

Gonorrhœa Injection.

This is prescribed by the most eminent medical practitioners in America.

℞ Sulphate Zinc,   -   -   15 grains.
    Fluid Ext. Golden-Seal,   -   4 drams.
    Listerine,   -   -   -   4 drams.
    Sulphate of Morphine,   -   5 grains.
    Soft or Rose Water to make 8 ounces.

DIRECTIONS:—Inject three or four times a day. All injections should be warm when used in any stage.

No. 36

Syphilis.

Internal Remedy.

R Carbonate of Iron, -   -   4 drams.
   Fluid Ext. Burdock, -   -   4 drams.
   Fluid Ext. Yellow-dock Root, 4 drams.
   Fluid Ext. Poke Root,   -   4 drams.
   Fluid Ext. Stillingia,   -   4 drams.
   Fluid Ext. Sarsaparilla,   -   4 drams.
   Simple Syrup to make -   6 ounces.

DOSE:—Teaspoonful four times a day.

This worst of all diseases can be successfully cured by this treatment.

No. 37

Syphilis.

External Application.

R Bluestone,   -   -   -   2 drams.
   Sugar of Lead, -   -   -   4 drams.
   Saltpetre,   -   -   -   4 drams.
   Dissolve in two pints of rain water.

DIRECTIONS:—Apply to chancres and sores three times a day. Apply dry Sub-Nitrate of Bismuth after using the above a few days.

No. 38

Chronic Gleet.

Internal Remedy.

Positive cure for this disgusting and noxious disease.

℞ Fluid Ext. Belladonna, - 1 dram.
Fluid Ext. Uva Ursi, - - 1 ounce.
Fluid Ext. Buchu, - - 1 ounce.
Fluid Ext. Cascara Sagrada, 1 ounce.
Syrup Rhubarb, - - 3 ounces.

DOSE:—Teaspoonful three times a day: at 9 a.m., 3 p.m., and 9 p.m.

No. 39

Chronic Gleet.

Injection.

℞ Boracic Acid, -   -   -   1 dram.
Glycerine, -   -   -   - 1 ounce.
Listerine,   -   -   -   2 ounces.
Sugar of Lead, -   -   - 20 grains.
Camphor Water,   -   -   6 ounces.

DIRECTIONS:—Use three times a day, warm. All injections should be warm when used in any stage.

No. 40

Loss of Vigor.

For Lost or Failing Vitality, Effects of Errors or Excesses in old or young. The most successful treatment known today.

℞ Tincture Cubebs, -   -   4 ounces.

DOSE:—A teaspoonful in a little water three times a day. May be continued for three or four weeks.

No. 41

Sores and Ulcers.

℞ Tincture of Iron.   Take internally.

DOSE:—Ten drops in a little water three times a day. Apply Powdered Magnesia to the sore, and bandage.

No. 42

Children Chafing.

℞ Powdered Starch, -   -   1½ ounces.
   Oxide of Zinc, -   -   -   3 drams.
   Powdered Camphor,   -   1 dram.
              Mix.

DIRECTIONS:—First thoroughly wash the parts with castile soap, and then apply as a powder.

No. 43

Glycerine Balm.

For Chapped Hands or Lips.

R   Glycerine, - - - 7½ ounces.
Tincture Myrrh, - - ¾ ounces.
Tincture Arnica, - - 1½ ounces.
Oil Rose, - - - 10 drops.

Rub Tincture of Myrrh with Carbonate Magnesia; add Water to make 9 ounces; then add Arnica and Glycerine.

DIRECTIONS:—Use three or four times a day.

No. 44

For the Complexion.

For Beautifying the Complexion.

R   Flake White, - - - 4 ounces.
Glycerine, - - - 4 ounces.
Bay Rum, - - - 2 ounces.
Rain Water, - - - 8 ounces.

Put in Carmine enough to color; dissolve Carmine in water; add Rose Water, 1 ounce.

Ladies will find this superior to all other preparations. It will not rub off, and makes the face and hands feel like velvet, and you know what you are using. Nothing injurious.

No. 45

Tooth-Powder.

This Powder is on the market today, has a large sale, and always gives satisfaction.

℞ Powdered Orris Root, -   4 drams.
    Powdered Peruvian Bark,   4 drams.
    Powdered Myrrh, -   -   4 drams.
    Prepared Chalk,   -   - 2 ounces.
    Oil of Cloves,   -   - 15 drops.

Mix thoroughly.

DIRECTIONS:—Use with brush as needed.

No. 46

Tooth-Powder.

℞ Powdered Cuttle Bone, -   1 ounce.
    Castile Soap, White, -   - 4 drams.
    Oris, Florentine,   -   - 4 drams.
    Oil Peppermint, -   -   - 20 drops.
    White Sugar, Powdered, -   1 ounce.

Rose Pink, if desired, to color.

DIRECTIONS:—Use with brush as often as needed.

No. 47
Condition Powder.

℞ Crude Antimony, - - 1 ounce.
Powdered Lobelia, - - 1 ounce.
Powdered Ginger, - - 3 ounces.
Flour of Sulphur, - - 3 ounces.
Powdered Bay Berry, - 1 ounce.
Cream Tartar, - - - 4 ounces.
Powdered Saltpetre, - 4 ounces.

DOSE:—One teaspoonful three times a day in feed.
Make your own powders, and not buy old stock.

No. 48
Chicken Cholera.
Used by the largest chicken breeders in the
United States.

℞ Sulphate Soda, - - 8 ounces.
Epsom Salts, - - - 8 ounces.
Powdered Sulphur, - 8 ounces.
Powdered Copperas, - - 8 ounces.
Powdered Mandrake Root, 4 ounces.
Red Pepper, - - - 4 ounces.
Mix.

DIRECTIONS:—Give in feed three times a day.

No. 49
Liniment for Man or Beast.
This has been in use for over forty years.

℞   Spirits of Ammonia,   -   1 ounce.
  Tincture of Chloroform,   -   1 ounce.
  Tincture of Opium,   -   1 ounce.
  Tincture of Capsicum,   -   1 ounce.
  Red Aconite,   -   -   -   1 ounce.
  Sapo Camphor Soap,   -   -   3 ounces.

DIRECTIONS:—To be applied freely externally two or three times a day.

No. 50
Horse Liniment.
White Liniment for Stock or Swelled Legs of Horses, Scratches, etc.

℞   Turpentine,   -   -   -   1 pint.
  Cider Vinegar,   -   -   -   1 pint.
  Kerosene Oil,   -   -   -   ½ pint.
  Beat up two Eggs and mix.

DIRECTIONS:—Wash legs clean, and apply it freely night and morning.

For horses and cattle there is no better liniment.

Vermin Exterminator.

℞ Dissolve Alum in hot water, making a very strong solution.

DIRECTIONS:—Apply to furniture or crevices in the wall with a paint-brush.

This is a sure destruction to all noxious vermin, and invaluable because easily obtained. It is, moreover, perfectly safe to use and leaves no trace behind. When you suspect moths have lodged in the boards and carpets, wet their edges with a strong solution; wherever it reaches them it is certain death.

No. 52

Erasive Fluid.

For Cleaning Clothing, Removing Grease Spots, etc.

℞ Spirits of Ammonia, - 1 ounce.
Alcohol, - - - - 1 ounce.
Sulph. Ether, - - - 1 ounce.
Spirits of Camphor, - - 8 drams.
Transparent Soap, - - 8 drams.
Water sufficient for - - 1 pint.

DIRECTIONS:—Lay a blotter or piece of cloth underneath garment to be cleaned and apply with a sponge. After removing spots, use sponge with clear water.

No. 53
Bed-Bugs.

Sure preventive against Bed-Bugs.

℞ Take Quicksilver and mix with Lard.

DIRECTIONS:— Take a feather, and apply underneath the slats of the bed in the spring and fall, and you will never see a bug in your house.

No. 54
Wood Preservative.

A man who has tried it says that wooden posts treated as follows, at a cost of two cents apiece, will last so long that the party adopting it will not live to see his posts decay.

DIRECTIONS:—Take boiled Linseed Oil and stir in Pulverized Charcoal to the consistency of paint, and put a coat over the timber.

No. 55
Mucilage.

Cold Glue or Mucilage.   None better in use.

℞ Gum Arabic,   -   -   6¼ ounces.
Sugar, Granulated, -   -   1½ ounces.
Water, -   -   -   -   7½ ounces.
          Dissolve.
Then add Acetic Acid,   -   1¼ ounces.

No. 56

To Clean Wall-Paper.

MAKING IT LOOK AS BRIGHT AS WHEN NEW.

Take a loaf of fresh rye bread; remove all of the crust and pour on ammonia, working it into a thick ball of paste. Take this in your hand and rub the paper. Be careful not to let the finger-nails scratch the paper.

No. 57

To Clean Clothes.

FRENCH METHOD OF CLEANING CLOTHES
OF ALL KINDS.

Put the clothes to be cleaned into a clean tub and cover them with gasoline; let them remain two hours, then take them from the tub and hang on a line. When dry, brush thoroughly and press. This method will remove grease and dirt of all kinds.

No. 38

Hectograph.

How to make Hectograph.

R Glycerine, - - - 1½ pounds.
Pulverized Glue, Best, - 8 ounces.

DIRECTIONS:—Dissolve the glue in one pint of hot water; when thoroughly melted add glycerine, and have them well mixed. Pour in graph, and set in cool and level place to harden.

Use violet hectograph ink.

www.ingramcontent.com/pod-product-compliance
Lightning Source LLC
Chambersburg PA
CBHW022033190326
41519CB00010B/1697